Bibliographic information published by the German National Library:

The German National Library lists this publication in the National Bibliography; detailed bibliographic data are available on the Internet at http://dnb.dnb.de .

Imprint:

Print and binding: Books on Demand GmbH, Norderstedt Germany
ISBN: 978-3-668-19291-1

This book at GRIN:

http://www.grin.com/en/e-book/318688/inductive-charging-as-a-range-extender-for-battery-electric-vehicles

Geoffrey Hancock, Nathan MacPherson, Ajay Patel

Inductive Charging as a Range Extender for Battery Electric Vehicles

Achieving highly efficient energy transfer through resonant inductive coupling between on-vehicle and in-road coils

GRIN Publishing

ABSTRACT

Adoption of electric vehicles has been hindered by the range anxiety due to relatively low energy density of current battery technology, and relatively long charge times. Current vehicles have supplemented electric capabilities by including a range extending internal combustion engine that allows for extra range when the level of charge of the battery is low. In order to maintain all-electric capability and achieve longer ranges, without improving battery capacity, a range extending charging system is needed. Inductive charging can be implemented for use while driving, so that the vehicle can be charged while it is moving. Resonant inductive coupling is a promising method for this application. Resonant inductive power transfer has been demonstrated to perform better than conventional inductive coupling, with larger gaps between source and receiving coils. This method has been demonstrated at high efficiency, 50% or more, at distances at least equal to the diameter of the source coil.

NOMENCLATURE

EV – Electric Vehicle

SOC – State-of-Charge

CS – Charge Sustaining

IC – Internal Combustion

RF – Radio Frequency

A2 – Ann Arbor

DTW – Detroit Metro airport

kWh – kiloWatt-hour

V – Volts

A – Amperes, or Amps

W – Watts

pF – pico-Farads

uH – micro-Henrys

MHz – Mega-Hertz

Contents

INTRODUCTION

PROBLEM DEFINITION AND SIGNIFICANCE With the current demand for more efficient vehicles on the rise, the desire for electric vehicles (EV's) with an extended range is increasing. Range anxiety is a huge issue with battery electric vehicles. Currently, many EV's utilize range extenders to quell range anxiety [1]. These on board generators however, rely on an internal combustion engine, to recharge the battery once battery reserve dips below allowable levels. Examples of these vehicles include the Chevrolet Volt and Fisker Karma [2]. Inductive wireless charging is a possibility to increase the range of these electric vehicles, but the problems associated with it, namely its low efficiency and slow charging rates, are holding the technology back [3].

INITIAL LITERATURE REVIEW Stationary inductive charging was successfully used by General Motors for the EV1 in the 1990's [4]. More recently, inductive charging has been introduced with smart phones and gaming systems, but the majority of consumer portable electronics still rely on wired charging [5]. Many patents have been published on using an inductive charging system to aid in electric vehicle propulsion, but they fail to actually map out a system and provide a tangible look at the feasibility of the technology [6, 7]. The patents lay out a scenario in which a vehicle would be charged inductively while traveling over roads equipped with conductors. The patents point out that the distance between the collector plates on the vehicle and the road conductors is critical to the efficiency of the overall system [8]. However, since the vehicle is moving this presents a challenge. If the plates are too close to the ground a small bump could dislodge the collectors, while too far away the system would not be efficient enough to be justified. Also, the design of the inductive charging system for highways has not been optimized for a specific vehicle or road.

OBSTACLES TO COMMERCIALIZATION Inductive charging for moving vehicles faces several challenges. As the air gap between conventional inductive coils increases efficiency decreases quickly. This poses a challenge of keeping the coil mounted on the moving vehicle very close to the ground without impact. A practical solution that maintains high efficiency is needed.

STATEMENT OF WORK Our goal is to design a range extending inductive charging system for vehicles driving on a highway. We will determine the relation between rate of charge for a certain design and the parameters related to operation including vehicle speed and gap distance between the charging system mounted on the vehicle and the road system. We will also determine the change in efficiency when those parameters are changed. We will design a system that could be applied to a Chevy Volt in order to maximize efficiency and rate of charge, with a goal of allowing the vehicle to drive for an extended range without using the range extending engine. For this we will need to determine the maximum rate at which its battery can charge (or the voltage it can handle), as well as physical dimensions. We will use the design characteristics of the induction system on the vehicle to design the inductive charging strips on the highway.

For a range extending inductive charging system, we will create a "Range Advisor" algorithm that will allow the driver to know how far he can go while charging or how often and how long he needs to be driving on the charging strip to maintain charge, or arrive at the highway exit with a desired state of charge. With our system designed we will apply the Range Advisor to a hypothetical duty cycle of an airport taxi, operating between the Ann Arbor (A2) area and the Detroit Metro Airport (DTW).

DELIVERABLES The following are the end of term deliverables:

1. Determine relation between rate of charge and:
    a. vehicle speed
    b. gap distance
2. Determine efficiency change due to variation of the same parameters
3. Design a theoretical system for an EV which maximizes rate of charge and efficiency of the inductive charger.
4. Perform laboratory experiment to prove feasibility of our system design
5. For this system a "Range Advisor" algorithm will advise how many trips can be made, for the duty cycle of an AA-DTW airport taxi.

FURTHER RESEARCH

EV CHARGING CHARACTERISTICS The battery that we are designing our system around is that of the Chevy Volt, and in order to proceed, we must understand the capacity of the Volt as well as its limitations. The theoretical capacity of the Volt battery is 16 kWh, a capacity that is never realized under normal conditions, usable capacity is about 12.8 kWh or 80% of the theoretical capacity. The approximately 40 mile all-electric driving range of the volt will drain 50% of the batteries maximum charge, bringing the volt battery level to 30% at 4.8 kWh. At 30% state of charge, the range extending IC engine supplements battery power in charge sustaining mode to stabilize the SOC at near 30% [9, 10].

In terms of the maximum rate at which users can charge the battery, the Volt battery can be charged at about 3.8 kW, or 240 V and 16 A, which corresponds to a C rate of approximately C/4, which will fully charge the battery in about 4 hours. At this rate of charge, the battery will last approximately 10 years/150,000 miles [11].

We also decided to look at charging the Nissan Leaf's battery for the sake of comparison. Nissan's battery has a maximum theoretical capacity of 24 kWh, and lasts approximately 8 years/100,000 miles. Onboard charging is limited by the 3.3 kW onboard charger, which provides a charge rate of about C/7. Nissan, however, provides customers with access to 480 volt, 125 amp chargers in a few of their dealerships, capable of a 2C charge rate (48 kW). The usable capacity is of the battery in the Leaf is also 80% of the theoretical capacity, or about 19.2 kWh. On one charge, the Nissan Leaf is capable of providing 100 mile all electric driving range [12].

Finally, we looked at how current batteries fared compared to the original EV-1 with the NiMH equipped batteries (not the original lead-acid batteries). These batteries were rated at 26.4 kWh, at 343 volts, and provided an all electric range between 75 and 150 miles per charge. To charge the 80% usable capacity completely, the charge rate is between C/3 and C, or 8.8 to 26.4 kW [13].

RESONANT INDUCTIVE CHARGING THEORY Power transfer through inductive coupling can be done through various methods. The efficiency of inductive coupling is greatly affected by the amount of magnetic flux that is "captured" by the secondary coil. The traditional way is to form what is basically a transformer without the high permeability core. In this method two circuits, one with the power source and one with the load, are connected via the electromagnetic coupling between an inductor on each circuit. The efficiency of this setup is greatly reduced as the air gap between the inductors increases; therefore this coupling is only practical at very small distances between primary and secondary coils. The circuit diagram for this setup can be seen in Figure 1(a) [14].

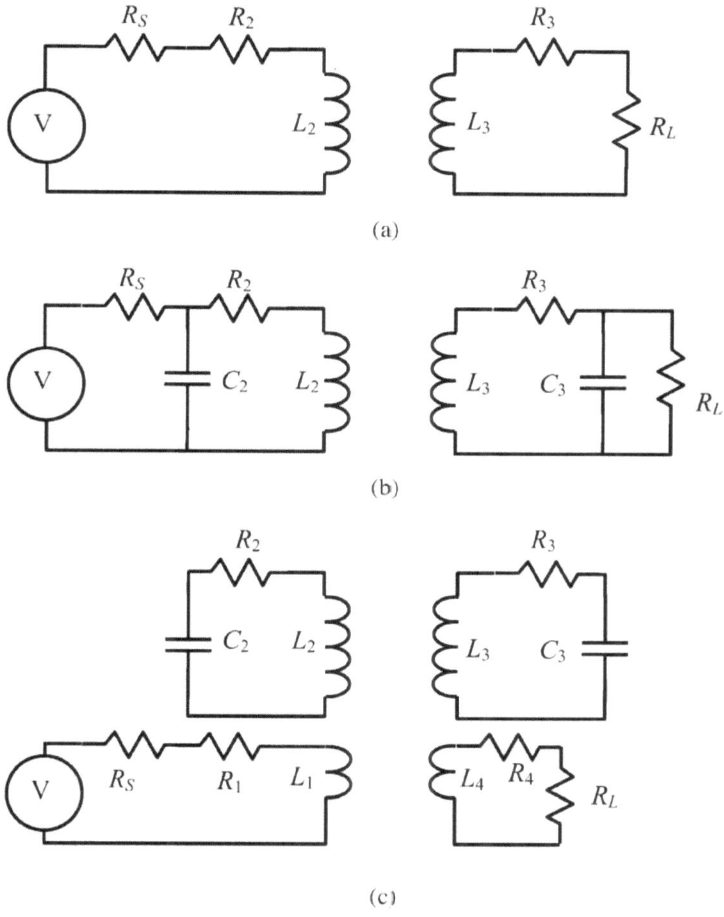

Figure 1: (a) Circuit diagram of inductive coupling between two coils. (b) Circuit diagram of resonant inductive coupling between two coils, with low Q. (c) Circuit diagram of resonant inductive coupling using four coils to achieve high Q.

One way to increase efficiency is to add capacitors to each circuit, as seen in Figure 1(b); this allows energy to oscillate back and forth, between an electromagnetic field in the inductor and a charge on the capacitor. This is called resonant inductive coupling. If the resonant frequency of the secondary circuit is tuned to the frequency of the primary circuit, high efficiency can be achieved with a larger gap between the two coils [15]. The limitation of this method is that Q, the quality factor, is low. A low Q means power is dissipated faster. Equation 1, equation for Q factor in an ideal series RLC circuit, shows that higher resistance in the coupled loops will decrease Q.

$$Q = \frac{1}{R}\sqrt{\frac{L}{C}} \quad (1)$$

In order to achieve a high Q, you can separate the resonating circuits so that each one has low resistance, as seen in Figure 1(c). Circuit 1 and 2, are coupled with coils of the same diameter, and circuits 3 and 4 are coupled with coils of the same diameter, circuits 2 and 3 can have different dimensions but are tuned to the same resonant frequency, ω0, by designing appropriate L and C values, as demonstrated by Equation 2 [15] .

$$\omega_0 = \frac{1}{\sqrt{L_2C_2}} = \frac{1}{\sqrt{L_3C_3}} \quad (2)$$

The mutual inductance, M, is a measure of the voltage induced in a secondary coil in response to a change in current in the primary coil, defined in Equation 3a. When mutual inductance is determined, it can be used to predict behavior of the secondary circuit, as in Equation 3b-3c. [16]

$$M_{ij} = k_{ij}\sqrt{L_iL_j} \quad (3a)$$

$$V_1 = L_1\frac{dI_1}{dt} - M_{12}\frac{dI_2}{dt} \quad (3b)$$

$$V_2 = M_{12}\frac{dI_1}{dt} - L_2\frac{dI_2}{dt} \quad (3c)$$

Mutual inductance is generally increased by keeping the secondary coil close to and oriented parallel with the primary coil [17]. The value "k" seen in equation 3 is the coefficient of coupling. This value indicates the strength of coupling between a pair of inductors, $0 \leq k \leq 1$; where a value of 1 is strongly coupled. When inductors are strongly coupled they behave like a transformer (i.e. low leakage flux), according to Equations 4(a) and 4(b), where voltage and current of the secondary coil is determined by a ratio of the number of loops in the coil, N.

$$V_s = \frac{N_s}{N_p}V_p \quad (4a)$$

$$I_s = \frac{N_p}{N_s}I_p \quad (4b)$$

PRELIMINARY SYSTEM DESIGN

One of the biggest problems facing inductive charging has been keeping a high efficiency with a large distance between inductors. By designing our system using resonant inductive coupling, we can achieve high efficiency, without worrying about keeping the secondary coil very close to the road. Another benefit of our preliminary design is that the secondary coil need not be the same diameter as the primary coil in the road. This means that the system will not require a large amount of space to be taken up on the underside of the vehicle.

THEORETICAL APPLICATION When applying inductive power transfer to in-road electric vehicle charging, one of the main constraints on the design is the distance between the underside of the car and the road. This distance is too large to achieve high efficiency through non-resonant inductive coupling, unless the secondary coil is suspended below the vehicle. Since resonant inductive coupling has demonstrated efficiency of more than 50% at distance larger than the diameter of the coils [9], this method seems to fit the application best.

Our system design is similar to the experiment designed by Cannon et al. [14]; a diagram of this system is shown in Figure 2. Another attractive feature of this design is that efficiency will not be significantly decreased if the loop diameters of coils 2 and 3 are different sizes. This allows a large loop in the road coil, and a smaller loop in the car coil.

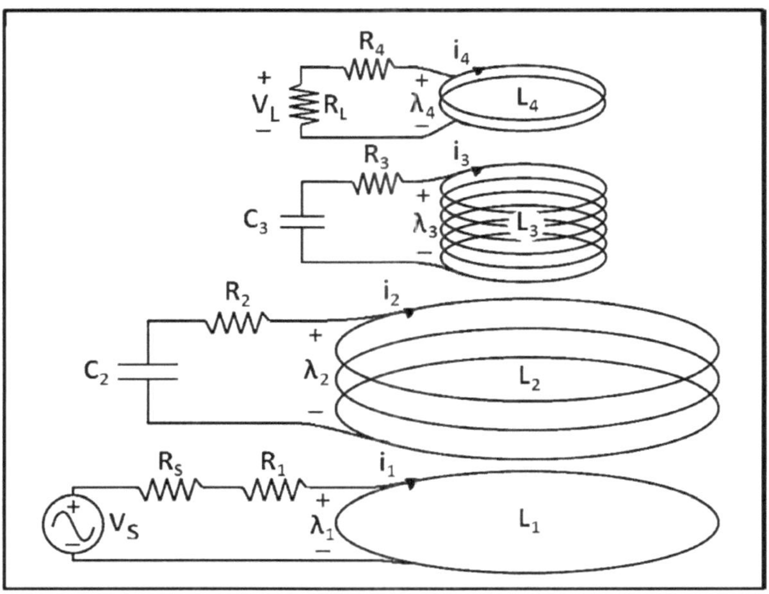

Figure 2: Preliminary Design: System Diagram

Using Kirchoff's voltage law around each of the four coils, equations 5a-5d are derived for the voltage in each coil. The rate of change of the flux linkage, λ, is used to describe the voltage at the terminals of the loop. Each of the coils is inductively coupled to the other three, so the flux linkages in each of the coils by the product of inductance and current, as seen in Equation 6 [14].

$$V_s = (R_s + R_1)I_1 + j\omega\lambda_1 \qquad (5a)$$

$$0 = R_2 I_2 + \frac{I_2}{j\omega C_2} + j\omega\lambda_2 \qquad (5b)$$

$$0 = R_3 I_3 + \frac{I_3}{j\omega C_3} + j\omega\lambda_3 \qquad (5c)$$

$$0 = (R_L + R_4)I_4 + j\omega\lambda_4 \qquad (5d)$$

$$\begin{bmatrix} \lambda_1 \\ \lambda_2 \\ \lambda_3 \\ \lambda_4 \end{bmatrix} = \begin{bmatrix} L_{11} & M_{12} & M_{13} & M_{14} \\ M_{12} & L_{22} & M_{23} & M_{24} \\ M_{13} & M_{23} & L_{33} & M_{34} \\ M_{14} & M_{24} & M_{34} & L_{44} \end{bmatrix} \begin{bmatrix} I_1 \\ I_2 \\ I_3 \\ I_4 \end{bmatrix} \qquad (6)$$

To find each of the mutual inductance values, all of the respective coupling coefficients must be determined. This can be done by estimating the fraction of magnetic flux that links each pair of coils, and can be confirmed through experimentation. Once each coupling coefficients are found the mutual inductances can be calculated using Equation 3a, as mentioned earlier.

Theoretically, there is no power transfer limit, even with a low coupling coefficient, if the system operates at the resonant frequency of the secondary coil [18].

METHODS

The intent of this work is to test the feasibility of in-road inductive charging for vehicles. Theoretical feasibility was tested experimentally, and feasibility of implementation was tested using simulation of an inductive charging system on real vehicles.

Resonant Inductive Coupling Experiment The purpose of this experiment was to prove that resonant inductive coupling would be feasible for use in vehicle charging, and that it provided a noticeable improvement of transferred voltage when compared to regular inductive coupling. The circuit tested, is similar to the design discussed previously in the Preliminary Design section.

Component Values for Resonance This experiment relies on two coils of wire (LC circuits) resonating at the same frequency. The components used to create these circuits must be chosen and tuned so that their frequencies match. All component values can be estimated, except for 1 capacitor in the second circuit, which must be a variable capacitor so its value can be tuned to produce the desired resonant frequency in the second circuit to match that of the first circuit. Resonant frequencies of the circuits were found by putting each coil in a circuit with a sinusoidal voltage input, and changing the sinusoidal input until a peak voltage was found. The circuit used is shown in Figure 3.

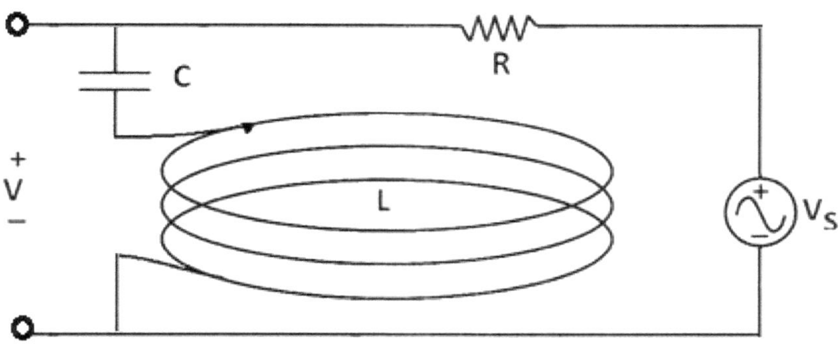

Figure 3: Circuit for testing resonant frequency of the LC coils.

EXPERIMENT SETUP/PROCEDURE For this experiment, we use four coils: a large 1 turn coil, 30 cm diameter, 10 gauge insulated wire that is driven by a sinusoidal input of 10 V at 2.4 MHz, a resonant coil of the same diameter and gauge with 3 turns (L = 8.2 uH) connected in series with 47 pF, a smaller 4 cm diameter resonant coil of 16 gauge wire with 3 turns (L = 0.82 uH) connected in series with a 470 pF capacitor that receives the charge, and a 1 turn 4 cm diameter 16 gauge coil that picks up the transferred charge.

The large driven coil is hooked up to the function generator, while the large capacitive coil is simply taped on top of the first coil. These two remain stationary for the experiment, and in order to make sure they are coupled at the highest efficiency, they must be as close to the same diameter as possible. The smaller capacitive coil is taped to the last small coil, again of the exact same diameter. This voltage in the last coil is measured with an oscilloscope in order to determine the charge being transferred to the pick-up loop. Experimental setup is shown in Figure 4.

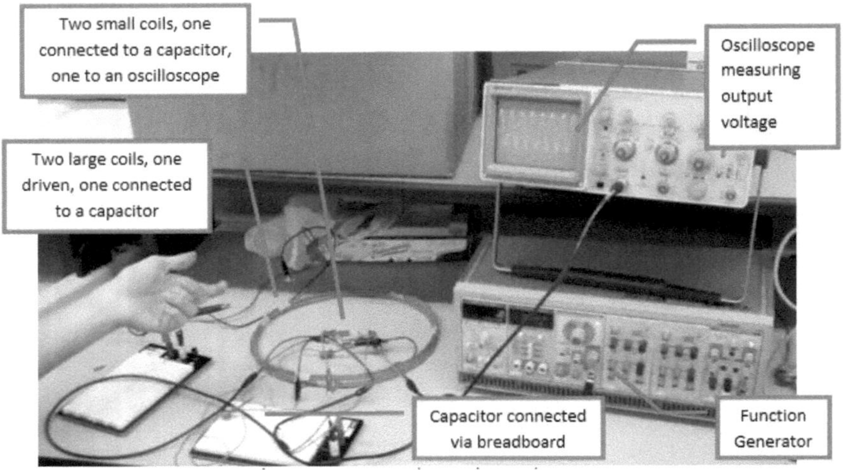

Figure 4: Resonant Inductive Coupling Experiment Setup

The first step of the experiment is to specify a constant AC voltage with which to provide the driven larger coil. This will be the comparison with which the efficiency of power transfer is determined. Once set, the two smaller coils, taped together, are to be set at the center of the large coils, and the function generator provides the sinusoidal input voltage matching the resonant frequency of the two LC circuit coils.

Now, observe the voltage measured in the last coil, measured with the oscilloscope. By moving the set of small coils, the effect of distance between coils, and movement of the small coils relative to the large coils can be quantified by the change in voltage.

To show the difference between resonant and normal inductive charging, simply remove the connection of the capacitors and repeat the experiment. Observe voltage at the final coil changes due to the same movement of small coils.

STATE OF CHARGE SIMULATION In order to calculate the parameters that the inductive charging system needs to possess a Simulink model was designed to simulate a Chevrolet Volt, Nissan Leaf and EV1 over a trip. The simulated trip is for a taxi driver who regularly takes University of Michigan students to and from the airport. The trip chosen was between Ann Arbor and Detroit Metropolitan Airport. The highway portion of the trip was modeled which was a total of 21 miles. The EPA Highway cycle was scaled to meet the top speed on a Michigan Highway, chosen to be 75 mph since the EPA data is for a highway with a speed limit of 55 and has a highest speed of 60 mph. [19].

The model starts with an input of time which feeds into the look-up table for the velocity at that second. The trip lasts a total of 1245 seconds, after which the cycle repeats until the vehicle's SOC is zero. The velocity then moves to the subsystem where tractive effort for the force of aerodynamic drag, rolling resistance and grade of road are calculated. It is assumed that the highway is relatively flat so a value of 0 degrees is assumed for the grade angle. Once the tractive effort is calculated, the model then calculates the energy required to travel on the highway at the input speed. The energy calculation is then used to find state of charge for the vehicle over time. The Simulink model for the SOC simulation is shown in Figure 5 and the subsystem model for calculating vehicle tractive power is shown in Figure 6.

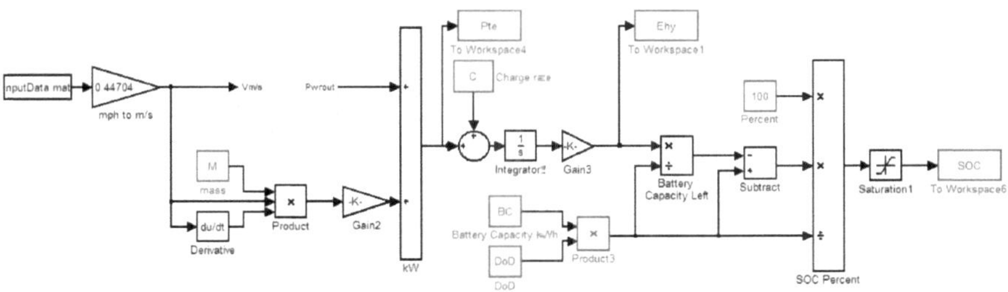

Figure 5: Main Simulink model for finding state of charge for a hybrid vehicle over time

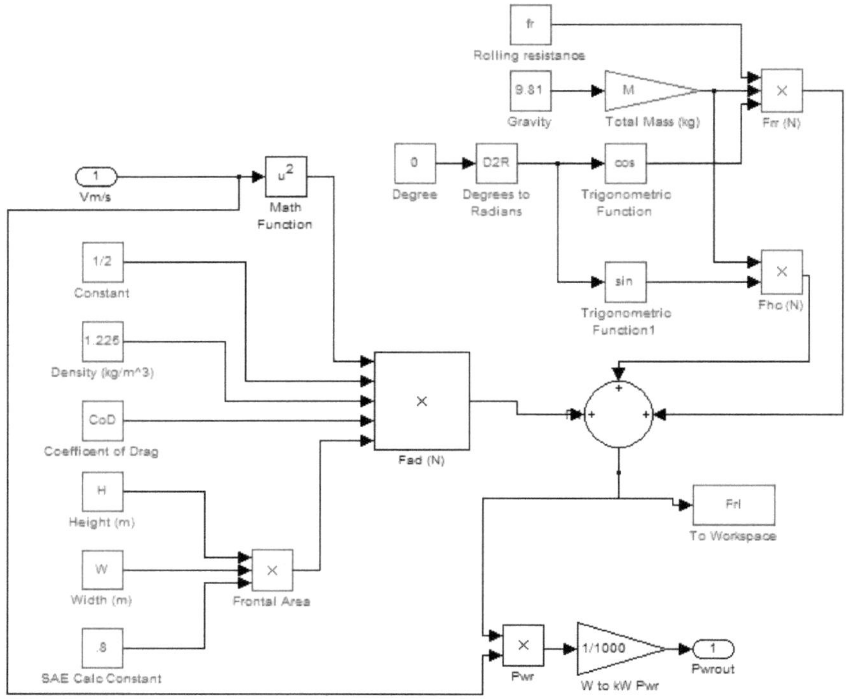

Figure 6: Simulink subsytem for calculating tractive power required over simulation time

Table 1 shows the inputs that were used in the model for each vehicle simulated. The depth of discharge is assumed to be 80% for the EV1 [20] and 50% for the Volt [21] and Leaf although exact figures for the newer models could not be referenced at the time of this report. The battery capacities and other vehicle parameters for the EV1 [22], Volt [23] and Leaf [24], except for rolling resistance were taken from manufacturer published data. The rolling resistance figure comes from the tire industry published figure for low rolling resistance tires which is standard equipment on all hybrids and electric vehicles [25].

Table 1: State-of-Charge Model Parameters

Input Parameters:	Volt	Leaf	EV1
Vehicle Mass [kg]	2,078	1,965	1,347
Coefficient of Drag	0.287	0.29	0.19
Vehicle Height [m]	1.43	1.55	1.28
Vehicle Width [m]	1.79	1.77	1.76
Coefficient of Rolling Resistance	0.09	0.09	0.09
Battery Capacity [kWh]	16	24	26.4
Depth of Discharge [%]	50%	50%	80%

RESULTS

RESONANT INDUCTIVE COUPLING EXPERIMENT Our experiment yielded results of 0.5 V of energy transfer to the small coil from an initial 10 volts provided to larger coil, for an efficiency of 5%.

A voltage of 0.5 V was measured in the final coil, this is the power transferred through the entire system. This voltage remained the same as the secondary coils were moved away from the primary coils, up to a distance of 30 cm. At that point the voltage started to decrease to a negligible amount at around 60 cm. Within 30 cm, moving the secondary coil relative to the primary coil did not decrease the output voltage.

The test with the capacitors removed, gave us results for the conventional inductive coupling experiment. This coupling yielded negligible energy transfer on the order of around 10 mV.

STATE OF CHARGE SIMULATION The simulations were run for each of the three vehicles and the SOC graphs are shown in the figures below. Different charging rates were looked at for each vehicle to see how many trips each could take before completely depleting the battery source.

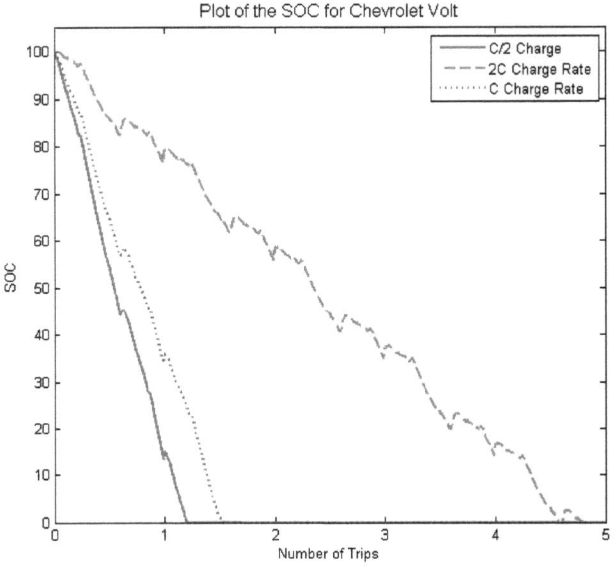

Figure 7: SOC graph for varying charge rates using the Volt.

Figure 7 shows the results of using the range advisor simulation over 2C, C and C/2 charge rates. When using the C and C/2 charge rates the vehicle is unable to continue in charge sustaining mode after just 1.5 trips. However, if the Volt batteries were able to be charged at a 2C rate then the vehicle is able to run for almost 5 total trips.

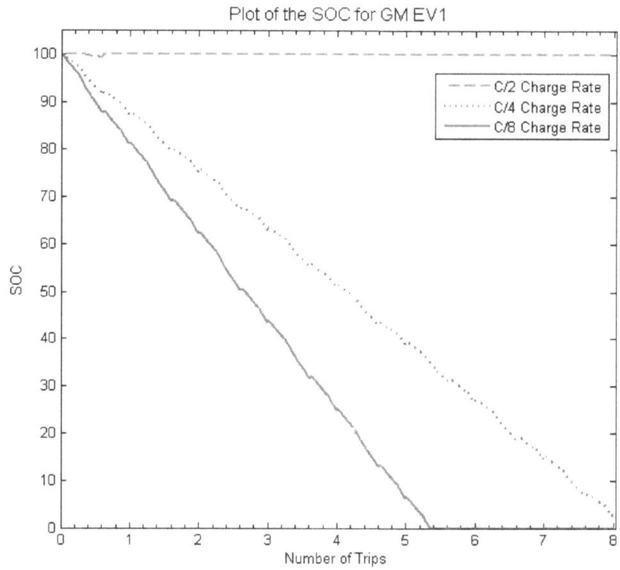

Figure 8 SOC graph for varying charge rates using the EV1.

Figure 8 shows the results of using the range advisor simulation over C/2, C/4 and C/8 charge rates. The EV1 was unique among the three vehicles in that anything above a C/2 charge rate would allow the vehicle to drive in perpetuity without losing any SOC. In order to see the lower bounds for the EV1 to compare it to the other vehicles lower charge rates of C/4 and C/8 were simulated. Even on the C/8 rate the EV1 still was able to run 5 trips while it took a 2C charge rate to do this same number of trips for the Volt.

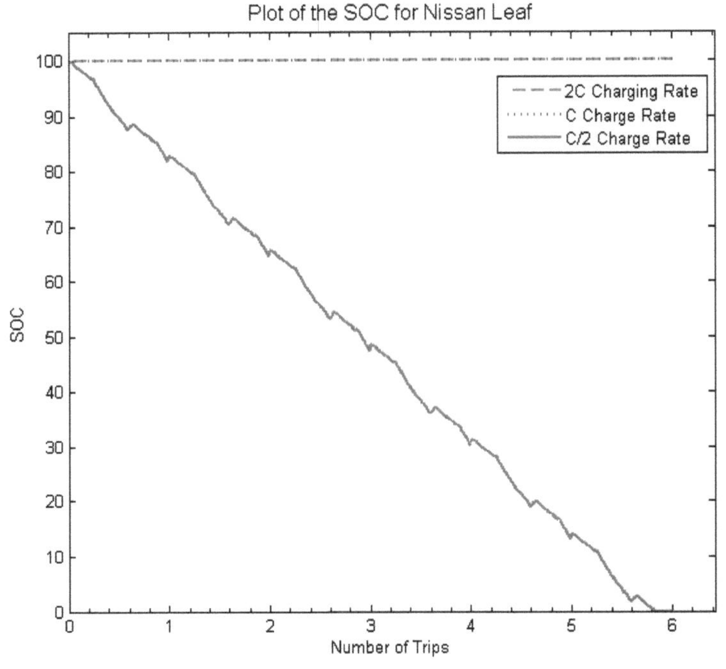

Figure 9 SOC graph for varying charge rates using the Leaf.

Figure 9 shows the results of using the range advisor simulation for the Nissan Leaf. The Leaf, like the EV1, would be able to continue making trips on any charge rates above C. A C/2 charge rate was simulated for comparison to the other two vehicles. The Nissan Leaf would be able to travel for 6 trips on the C/2 charge rate in comparison to the Volt which would only be able to make 1 trip.

DISCUSSION

RESONANT INDUCTIVE COUPLING EXPERIMENT Our experiment did not achieve the efficiency that we expected. This is not because the theory is not feasible. This efficiency is so low because we were not able to achieve matching frequencies in our two resonant coils. The coils were not able to be tuned to the exact resonant frequency due to the resolution limitations of our function generator, and we did not have a variable capacitor and thus were stuck with a constant capacitance. Instead of varying capacitance we attempted to vary inductance by changing the number of loops in the coil, but this was did not give us the fine tuning required. Due to these inadequacies, the result was decreased power transfer between coils.

We did, however, learn that there is a distinctly large difference between resonant and normal inductive charging. All of our inefficiencies were kept constant between the two tests and therefore allowed for an accurate comparison. The conventional inductive coupling yielded much lower power transfer than the resonant inductive coupling system. This justified our decision to pursue resonant inductive charging as opposed to normal inductive charging.

Similar experiments have achieved efficiencies of at least 50% at distances equal to the diameter of the primary coil, at small distances power transfer can be up to 90% [9,10]. Based on our experiment and other recent experiments, we have determined that it is feasible to use resonant inductive coupling for an inductive charging system for charging EVs. The power transfer efficiency of resonant inductive coupling is much greater than conventional inductive coupling, and still maintains good efficiency at much larger distances than allowed by conventional inductive coupling.

STATE OF CHARGE SIMULATION The range advisor simulations showed that the SOC for each of the vehicles was heavily dependent on charge rates selected based on the charge acceptance limitations on the battery pack. It is clear that current charge rates will be able to accommodate vehicle trips without having to stock to 'plug-in' but the feasibility of such a high power system, given the losses innate in inductive charging may be prohibitive to the technology.

The limitations of this simulation come from the difference between ideal charge situations and actual real world limitations of vehicle charging. Since inductive charging relies so heavily on the distance between the vehicle and the charging source, any variance in this would need to be address in future models and presents an engineering challenge to designers. The ideal would be to have the charge collector just above the charging plate but physical limitations and real world vehicle dynamics make this impossible.

Other factors which the model does not currently address are battery limitations due to factors such as ambient temperature and life of the pack. At very low temperatures the battery will not be as efficient and the SOC will degrade faster over the trip when compared to more favorable conditions. The life of the pack may also play a role as the battery pack degrades it may be less able to charge at the same rates as it could in a new state. The model in this report only focused on new vehicles in ideal conditions.

FURTHER DESIGN CONSIDERATIONS

SYSTEM DESIGN We have determined that at least 50% efficiency of power transfer can be achieved at a distance up to the diameter of the primary coil, using resonant inductive coupling. A diagram of our system design is shown in Figure 10.

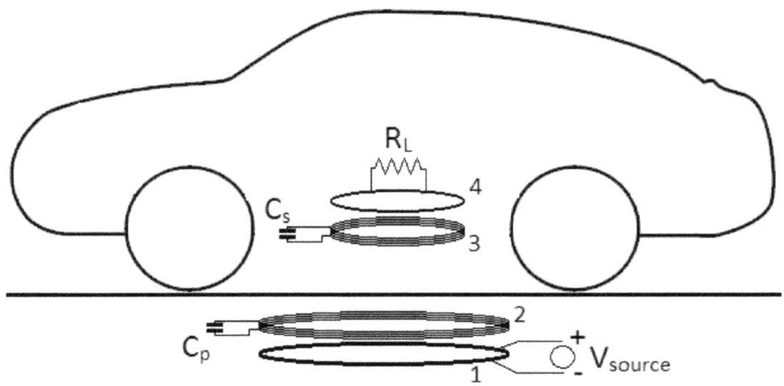

Figure 10: Preliminary Design of In-Road Inductive Charging System

This system is designed using the Chevrolet Volt as a reference. We estimate the gap between the underside of the vehicle and the road, or between coils 2 and 3, to be about 30 cm (1 foot). This means the coils in the road should be at least 30 cm. Coils on the underside of the car will be from 10cm up to 30 cm in diameter. There will be no significant gap between coils 1 and 2, or between coils 3 and 4, these couplings transfer power through conventional inductive coupling. By using resonant coupling between intermediate coils (coils 2 and 3) which have very low resistance (the source and load resistances are in coils 1 and 4), the coupling coefficient and quality factor are maintained at a high level. This is the reason for having 4 coils, with the two intermediate coils resonantly coupled, rather than just 2 coils resonantly coupled.

The Chevy Volt can be charged at 240 V (AC), and 3.3 kW. This means the source voltage should be at least 480 V for the same current, therefore requiring at least 6.6 kW. This will allow the battery to be charged at a rate of about C/4. The Nissan Leaf will have the capability of charging at 480 V and 125A (60 kW) in the near future, a rate of about 2C. Many other EVs will probably have this capability as well; therefore the system should be designed with that in mind, to accommodate future high rate charging capability. This would require an input power of about 120 kW at 50% efficiency.

SYSTEM APPLICATION This system can be applied for use on the highway, or for stationary use. For highway use, multiple source coils will be installed in the road along the desired stretch of road. Radio frequency (RF) technology, such as that used in open road tolling systems, can be used for sensing the cars approach as well as billing for using the charging system. When the car approaches a coil that coil will be activated, and when no cars are above the coils they can be deactivated. For stationary use a single source coil can be installed at stoplights, in parking spaces, or in garages.

CONCLUSION

Adoption of electric vehicles has been hindered by the range anxiety due to relatively low energy density of current battery technology, and relatively long charge times. Current vehicles have supplemented electric capabilities by including a range extending IC engine that allows for extra range when the level of charge of the battery is low. In order to maintain all-electric capability and achieve longer ranges, without improving battery capacity, a range extending charging system is needed. Inductive charging can be implemented for use while driving, so that the vehicle can be charged while it is moving. Resonant inductive coupling is a promising method for this application. Resonant inductive power transfer has been demonstrated to perform better than conventional inductive coupling, with larger gaps between source and receiving coils. This method has been demonstrated at high efficiency, 50% or more, at distances at least equal to the diameter of the source coil.

The system we have designed uses resonant inductive coupling, and can be used as a range extender for electric vehicles while driving on the highway. This system can also be used in stationary applications where conventional charging is not possible such as at stoplights or in parking spaces.

FUTURE WORK

To improve the results of the experiment attempted above, one factor that must be introduced is the ability to tune the resonant frequencies for matching; the above experiment dealt with constant capacitors and a low resolution function generator, which restricted the tuning abilities available, and thus diminished the efficiency of the system. Adding a variable capacitor and using newer, higher resolution equipment would yield more accurate tests and better results. With these experimental improvements in place, investigation should be done to determine the amount of power that can be transferred via resonant inductive coupling at the speed of a moving vehicle, relative to the source coil.

The next step of this research is to begin scaling upwards, gradually reaching real world testing situations. The resonant inductive charging experiment described earlier was conducted in controlled conditions with inaccurate measurement tools, low voltages, and imperfect materials with respect to wire length and diameter. The eventual final goal is for this system to be implemented on highways; therefore, the goal of testing is to reach high voltage transfer efficiency under appropriate weather conditions and vehicle load as well as highway speeds. The coils, voltage source, and overall circuit must be designed to handle the constant transfer of voltage from the ground to the vehicle. The goal is to perform many experiments to bridge the gap between the final goal and the experiment described earlier in this paper, each of which will tackle the variables mentioned above.

Cost of installing the system will likely be the biggest roadblock to implementation of this system. The possibility of retrofitting the system into existing roadways should be investigated. This could be done by creating channels in the road for inserting the coils, or designing thin and durable coils to be placed on the surface of the road.

REFERENCES

[1] E. D. Tate, Michael O. Harpster, M. O., Savagian, P. J. and Tate, E. D., 2008, "The Electrification of the Automobile: From Conventional Hybrid, to Plug-in Hybrids, to Extended-Range Electric Vehicles," SAE International, 200-01-0468, General Motors Corportation, Detroit, MI.

[2] Alam, M., Green, R. C., Wang, L., 2010, "The Impact of Plug-in Hybrid Electric Vehicles on Distribution Networks: A Review and Outlook," Renewable and Sustainable Energy Reviews, Article RSER-1078, University of Toledo, Toledo, Ohio.

[3] Waffenschmidt, E., Staring, T., 2009, "Limitation of Inductive Power Transfer for Consumer Applications," IEEE Power Electronics and Applications, 2009. EPE '09. 13th European Conference. Aachen, Germany.

[4] Hayes, J.G., 1998, "Battery Charging Systems for Electric Vehicles," IEEE Electric Vehicles - A Technology Roadmap for the Future (Digest No. 1998/262), London, UK.

[5] Yuan, Tong et al. "Advanced Inductive Charging Pad" Fresh Patents. Feb 18 2010 <http://www.freshpatents.com/-dt20100218ptan20100039066.php>

[6] Schwind, J. P., 1998, "Armature Induction Charging of Moving Electric Vehicle Batteries," United States Patent, 5,821,728, Jessup, Maryland.

[7] Dahl, E. A., 1976, "Induction Charging System," United States Patent, 3,938,018, Ventura, California.

[8] Bolger, J. C., 1975, "Supplying Power to Vehicles," United States Patent, 3,914,562, Orinda, California.

[9] Matthe, R. "From Vehicle Requirements to Li-Ion Cell Requirements." ESE-505: Energy Generatino and Storage Using Modern Materials. The University of Michigan, Ann Arbor, MI. 23 Nov. 2010. Lecture.

[10] Dennis, Lyle. "Latest Chevy Volt Battery Pack and Generator Details and Clarifications," GM-Volt. Aug 29 2007 < http://gm-volt.com/2007/08/29/latest-chevy-volt-battery-pack-and-generator-details-and-clarifications/>

[11] Dennis, Lyle. “GM Announced Chevrolet Volt 240 V Charger Pricing and Installation Service Provider,” GM-Volt. Oct 10 2010 < http://gm-volt.com/2010/10/06/gm-announces-chevrolet-volt-240v-charger-pricing-and-installation-service-provider/>

[12] van Agt, Jeroen. “Electric Car Nissan LEAF,” Olino. Aug 2 2008 <http://www.olino.org/us/articles/2010/08/02/electric-car-nissan-leaf>

[13] Smalls, Robert. “EV1 Brochures and Analysis,” Eco-Modder. Aug 8 2010 <http://ecomodder.com/forum/showthread.php/ev1-brochures-analysis-14161.html>

[14] Cannon, B.L.; Hoburg, J.F.; Stancil, D.D.; Goldstein, S.C.; , "Magnetic Resonant Coupling As a Potential Means for Wireless Power Transfer to Multiple Small Receivers," Power Electronics, IEEE Transactions on , 24, no.7, pp.1819-1825, July 2009

[15] A. Kurs , A. Karalis , R. Moffatt , J. D. Joannopoulos , P. Fisher and M. Soljacic "Wireless power transfer via strongly coupled magnetic resonances", Science, 317, pp. 83 2007.

[16] Chwei-Sen Wang; Stielau, O.H.; Covic, G.A.; , "Design considerations for a contactless electric vehicle battery charger," Industrial Electronics, IEEE Transactions on , 52, no.5, pp. 1308- 1314, Oct. 2005

[17] Grover, Frederick W. *Inductance Calculations*. New York: D. Van Nostrand, 1946. Print.

[18] Moradewicz, A. J.; Kazmierkowski, M. P. “High efficiency contactless energy transfer system with power electronic resonant converter”, BULLETIN OF THE POLISH ACADEMY OF SCIENCES-TECHNICAL SCIENCES, 57, no.4, pp. 375-381, Dec 2009.

[19] EPA. "Detailed Test Information." *Fuel Economy*. 2008. Web. 20 Nov. 2010. <http://www.fueleconomy.gov/feg/fe_test_schedules.shtml>.

[20]Brost, Ron D. *Performance of Valve-Regulated Lead Acid Batteries in EVI Extended Series Strings*. Rep. no. 0-7803-4098-1/98. Indianapolis: Delphi Automotive Systems, 1998. IEEE, Jan. 1998. Web. 20 Nov. 2010. <http://ieeexplore.ieee.org/stamp/stamp.jsp?tp=&arnumber=653833&userType=inst>.

[21]Brooker, Aaron, Matthew Thornton, and John Rugh. "Technology Improvement Pathways to Cost-effective Vehicle Electrification." *Technology Improvement Pathways to Cost-effective Vehicle Electrification* (2010). *SAE Technical Papers*. SAE International, 12 Apr. 2010. Web. 20 Nov. 2010. <http://papers.sae.org/2010-01-0824/>.

[22]Electric Transportation Applications. "General Motors EV1 Vehicle Specifications." Department of Energy, 1996. Web. 20 Nov. 2010. <avt.inel.gov/pdf/fsev/eva/genmot.pdf>.

[23]"Chevy Volt Specs | GM-VOLT : Chevy Volt Electric Car Site." *GM-Volt: Chevy Volt Electric Car Site*. Web. 20 Nov. 2010. <http://gm-volt.com/full-specifications/>.

[24]Nissan. "Nissan Leaf." *Nissan Leaf Specifications*. Web. 20 Nov. 2010. <http://www.nissan-zeroemission.com/EN/LEAF/specs.html>.

[25] Transportation Research Board. "Tires and Passenger Vehicle Fuel Economy: Informing Consumers, Improving Performance -- Special Report 286." *The National Academies Press*. 2006. Web. 20 Nov. 2010. <http://www.nap.edu/catalog/11620.html>.